iMath
Readers

Space Adventures:
Where Does the Time Go?

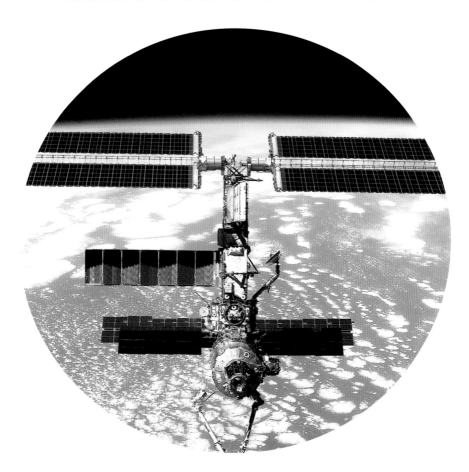

by John Perritano

Content Consultant
David T. Hughes
Mathematics Curriculum Specialist

NORWOOD HOUSE PRESS
Chicago, IL

Norwood House Press
PO Box 316598
Chicago, IL 60631

For information regarding Norwood House Press, please visit our website at
www.norwoodhousepress.com or call 866-565-2900.

Special thanks to: Heidi Doyle
Production Management: Six Red Marbles
Editors: Linda Bullock and Kendra Muntz
Printed in Heshan City, Guangdong, China. 208N—012013

Library of Congress Cataloging–in-Publication Data

Perritano, John.

  Space adventures: where does the time go? / by John Perritano; content
  consultant, David Hughes, mathematics curriculum specialist.
  pages cm.—(iMath)

  Audience: 8–10
  Audience: Grade 4 to 6
  Summary: "The mathematical concepts of elapsed time, converting units
  of time, and equivalent units of time are introduced as readers learn about
  astronauts aboard the International Space Station. Concepts include adding
  and subtracting units of time, using a number line diagram, and time
  intervals. This book features a discover activity, connections to science
  and history, and mathematical vocabulary introduction"—Provided
  by publisher.

Includes bibliographical references and index.

ISBN: 978-1-59953-562-3 (library edition: alk. paper)
ISBN: 978-1-60357-531-7 (ebook) (print)

1. Time measurements—Juvenile literature.
2. International Space Station—Juvenile literature.
3. Space and time—Juvenile literature. I. Title.

QB213.P47 2012
529—dc23
2012034234

CONTENTS

Note to Caregivers:

Throughout this book, many questions are posed to the reader. Some are open-ended and ask what the reader thinks. Discuss these questions with your child and guide him or her in thinking through the possible answers and outcomes. There are also questions posed which have a specific answer. Encourage your child to read through the text to determine the correct answer. Most importantly, encourage answers grounded in reality while also allowing imaginations to soar. Information to help support you as you share the book with your child is provided in the back in the **Additional Notes** section.

Bold words are defined in the glossary in the back of the book.

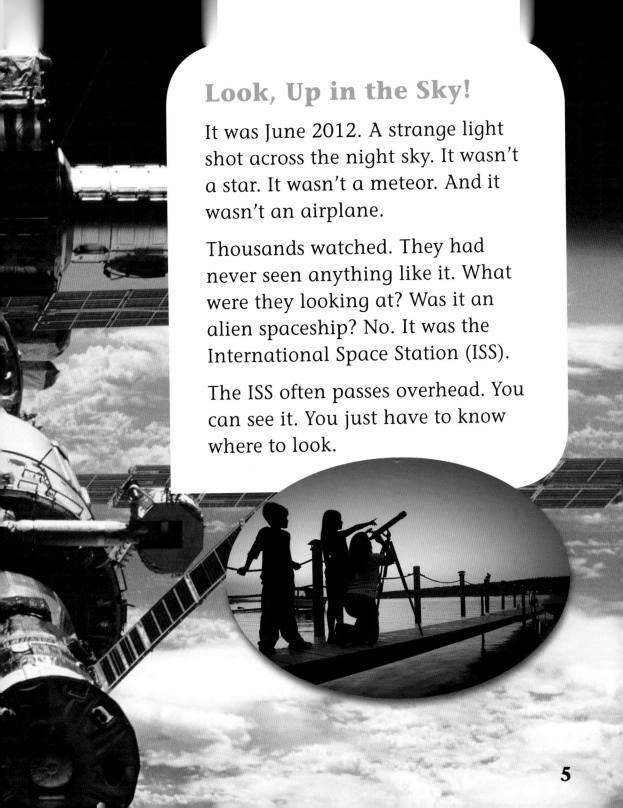

Look, Up in the Sky!

It was June 2012. A strange light shot across the night sky. It wasn't a star. It wasn't a meteor. And it wasn't an airplane.

Thousands watched. They had never seen anything like it. What were they looking at? Was it an alien spaceship? No. It was the International Space Station (ISS).

The ISS often passes overhead. You can see it. You just have to know where to look.

The ISS is more than a rapid space machine. It's a home and office. It's a library and a place for fun and games. Humans have long dreamed of living in space. The ISS brings that dream closer.

The ISS launched in 1998. It is about 200 miles above Earth. It **orbits**, or circles our planet. It takes about 90 minutes for the ISS to orbit Earth once. It moves at 17,000 miles per hour.

The ISS turned 10 years old in 2010. By then, it had traveled 1.5 billion miles. It had orbited Earth more than 57,000 times.

The first crew arrived at the ISS in 2000. Since then, the space station has been a home away from home. Astronauts from all over the world have lived there.

Astronauts are space scientists. They work, eat, and play aboard the ISS. They stay on the ISS for months. They follow strict **schedules**. Schedules are timetables. In this book, you will learn more about time.

Many people dream of living in cities in outer space. This illustration shows what a space city might look like.

Sometimes, you want to know how much time has passed. Time that has passed is called **elapsed time**.

Time passes in units. Those units include seconds, minutes, hours, and days. You may need to know how to **convert**, or change, units of time to find elapsed time. Look at the chart. It shows **equivalent**, or equal, amounts of time.

Equivalent Units of Time

60 sec.	1 min.
15 min.	$\frac{1}{4}$ hr.
30 min.	$\frac{2}{4}$ hr., or $\frac{1}{2}$ hr.
45 min.	$\frac{3}{4}$ hr.
60 min.	1 hr.
24 hr.	1 d.

sec. = second or seconds
min. = minute or minutes
hr. = hour or hours
d. = day or days

There are different ways to find elapsed time.

Idea 1: Move the hands of a clock to show elapsed time. Say the starting time for an event is 2:20 P.M. Put the hands at the starting time. Then, move the hands to the ending time. Count the minutes and hours as you move the hands.

Idea 2: Write and solve a subtraction problem to find elapsed time. Say an astronaut worked on an experiment from 1:50 P.M. to 4:45 P.M. How much time elapsed? Set up a subtraction problem, using a **regrouping chart**. Regroup to subtract minutes. Then, subtract hours.

	Hours	Minutes
Finish Time	3 4̶	60 + 45 = 105 4̶5̶
Start Time	1	50
Elapsed Time	2	55

Idea 3: Use a **number line diagram** to show elapsed time. Say an astronaut had to fix a problem on the outside of the ISS. She stepped outside at 11:50 A.M. It took her $4\frac{1}{2}$ hours to fix the problem. What time did she come back into the station?

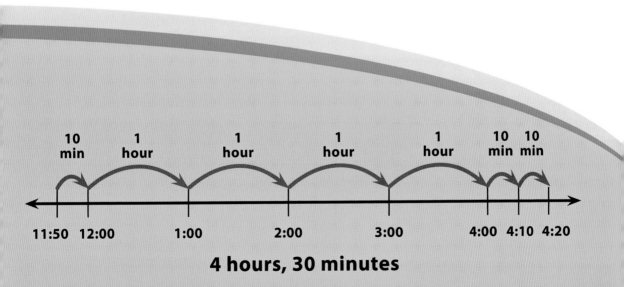

4 hours, 30 minutes

DISCOVER ACTIVITY

Materials
- watch or clock with a second hand
- a radio, computer, iPod, or MP3 player
- paper and pencil

Time That Tune

What is your favorite way to listen to music? Do you listen to CDs? The radio? Do you use an iPod or MP3 player? Or perhaps you listen to music on a computer. It doesn't matter which tool you use to enjoy music.

Draw a chart like this one:

Song Title	Start Time	End Time	Elapsed Time

Think about four songs you enjoy hearing often. Write the names of the songs in the chart.

Listen to each song. Use a clock or watch to time the songs. Write the song's start time and end time in the chart.

When you have finished, think about how you will find the elapsed time of each song. Will you

- move the hands of a clock?
- write and solve a subtraction problem?
- use a number line diagram?

Find the total amount of elapsed time for all four songs.

How much time elapsed while you were listening to music?

Go Team!

The ISS is the largest human-made object in space. It's as large as a football field. It weighs nearly one million pounds. That's the weight of more than 320 automobiles! The station is powered by **solar cells**. They stick out from the sides of the station like wings. The cells turn energy from the sun into electricity.

 Did You Know?

The U.S. had a space station before the ISS. It was known as *Skylab*. *Skylab* orbited Earth for more than five years.

The solar cells on the ISS power the spacecraft.

Building something so big requires teamwork. Beginning in 1998, workers from 16 countries helped build the ISS. Those countries included the United States, Canada, Russia, Brazil, and Japan. There were also 11 countries in the European Space Agency.

Putting the ISS together was like building a model race car. Each piece had to fit. It wasn't an easy job. Astronauts had to assemble the ISS in space.

Astronauts blasted into space on the **space shuttle**, a space ship. Astronauts on the shuttle worked above Earth. Building the ISS was not easy. Workers built the ISS in stages. They used cranes to put parts together. They used a robot camera to see what they were doing. Sometimes, astronauts walked in space to work on the station.

Imagine yourself working in space to help build the ISS.

? Did You Know?

It took 162 space walks totaling 1,021 hours to assemble the ISS. How many days is that?

The chart below shows your trips outside the station. Copy the chart. Then, fill in the missing information.

Space Walk	Start Time	End Time	Elapsed Time
No. 1	8:15 A.M	9:30 A.M	?
No. 2	?	12:00 P.M.	1 hour 30 minutes
No. 3	?	4:20 P.M.	2 hours 15 minutes
No. 4	2:20 A.M.	7:15 A.M	?
No. 5	4:15 P.M.	?	1 hour and 20 minutes

What happens if you drop an apple? It falls to the floor. Why? **Gravity** makes it happen.

Gravity is a kind of **force**. It acts on objects with **mass**. The apple has mass. So does Earth. Gravity attracts the apple toward Earth's center.

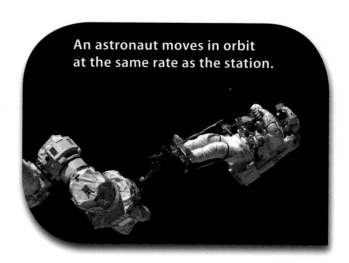

An astronaut moves in orbit at the same rate as the station.

There is also gravity between the ISS and Earth. That's because Earth has much more mass than the ISS. Gravity works inside the ISS, too. If an astronaut drops an apple, the apple falls. But it doesn't fall toward Earth. It falls around Earth. That's because Earth's gravity keeps objects circling the planet. It's the same force that keeps Earth in orbit around the sun.

The station, the astronauts, and the apple fall around Earth at the same rate. So, they seem to float, as though they have no weight. This floating state is called **microgravity**.

Some astronauts do research aboard the ISS. They study the effects of microgravity. Some study how cells and **proteins** grow in space. Proteins are the building blocks of life. They are found in the cells that make up all living things. The proteins in a human body do different jobs. Some make our hair, muscles, and bones. Others carry oxygen in our blood.

Companies that make medicine want to know how proteins work. So, they study how proteins grow in space.

Imagine that a scientist did three experiments. She wrote the times in a chart. How much time did she work in all?

	Hours	Minutes
Experiment 1	4	20
Experiment 2	5	45
Experiment 3	2	30

Connecting to Science

Santa Clara is a town in California. Some 8th-grade students there are part of a Student Spaceflight Experiments Program.

The students designed a microgravity test that they entered into a contest. Their design won. That means that astronauts will carry out their experiment aboard the ISS.

The experiment will study a kind of plastic clay. The rubbery material stretches and bounces. The experiment will determine if the toy works the same way in space as it does on Earth.

Students came up with an experiment to see if plastic clay is still rubbery in outer space.

The students had worked hard to win. They were overjoyed!

Imagine that the students worked for several weeks to design their experiment. They wrote their times for one week in the chart below. How much time did they work this week?

Day	Hours	Minutes
1	2	25
2	1	45
3	0	50
4	1	15
5	1	55

Other winning experiments included one from fifth-grade students. They wanted to know how the sprays people use to kill **mold**, or fungus, work in space.

Mold grows in a dish in a lab.

Life on the ISS

There aren't any grocery stores in outer space. There are no fast food restaurants. Astronauts can't order pizza. The closest hot dog stand is about 220 miles away. That's how far away the ISS is from Earth most of the time. Still, astronauts have to eat.

Luckily, a delivery truck flies food to the ISS. It's not a real delivery truck. It's a spaceship. Every few months a spaceship travels to the ISS. It's loaded with goodies. There's fresh fruit. There's water. There are packaged meals. There are even tasty desserts.

It's time for dinner aboard the ISS.

There is a kitchen table on the ISS. The table has straps and Velcro to keep food containers from floating away.

There are no kitchen chairs. Instead, there are metal bars below the table. Astronauts put their feet beneath the bars and stand up to eat.

The ISS menu has about 100 items. Astronauts pick their favorite foods before each flight. They have ketchup, mustard, and mayonnaise, too. There's even salt and pepper.

Salt and pepper are liquids aboard the ISS. Why is that? Grains of salt or pepper float in space. Astronauts don't want the grains to break any equipment.

Most food is **dehydrated**. That means the liquid in it has been removed. So, astronauts put the liquid back in.

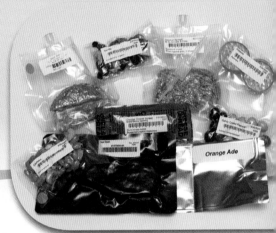

Astronauts have many foods to choose from.

Orange Ade

Even tea, coffee, milk, and juices are dehydrated. Each bag has a tube on one end and a straw on the other. Astronauts cut the tube end and pour water into the bag. Then, they mix the drink. Next, they cut the straw end. They suck the drink through the straw.

Say that ISS astronauts stopped work at 3:15 P.M. They ate dinner 1 hour and 20 minutes later. What time did the astronauts begin eating?

Exercise Time

Long space flights change an astronaut's body. For example, on Earth, a human's heart squeezes to push blood out to the body. The blood travels to the head and neck. And gravity helps it reach the lower body.

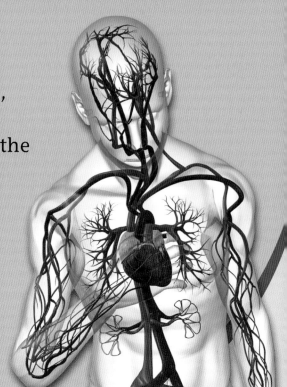

In space, an astronaut's heart works like it does on Earth. But there is only microgravity. So, most of the blood goes to the head and neck. The result is a puffy face. Blood vessels to the head and neck are filled with blood. The eyes turn red. Vessels along the neck swell.

The opposite happens to an astronaut's legs. Without help from gravity, only the heart works to get blood to the lower body. The result is skinny legs.

Doctors on the ground watch each astronaut's health carefully. Good health includes time for exercise.

The ISS has a gym. The gym has a **treadmill**. A treadmill is a machine with a belt. A person can set the rate at which the belt moves. That allows the person to run or walk. The treadmill on the ISS floats like the astronauts do. An astronaut who wants to run or walk on the machine gets into a set of belts first. Otherwise, the astronaut could send the machine floating away with only a footstep.

There is also a cycling machine bolted to the floor. Astronauts strap their shoes into buckles and wear a seat belt so they don't float away.

In this photo, Astronaut Edward T. Lu, the Expedition 7 NASA ISS science officer and flight engineer, exercises on the treadmill.

Astronauts use the gym's **resistance** machine to lift weights so their muscles don't get weak.

Imagine that an astronaut rode the treadmill from 12:00 P.M. to 12:30 P.M. He spent one hour on the cycle. And he spent 15 minutes using the resistance machine. What time did his workout end? How much time did he exercise in all?

Time to Relax

Life on the ISS isn't always about work and exercise. People take time to relax. They send and receive email from home. They play games.

Astronauts take all sorts of things into space. Some take checkers or chess sets. Others bring musical instruments. They take books to read. They take CDs to listen to, and DVDs to watch.

Astronaut Carl E. Walz entertains his crewmates.

Imagine members of a crew taking turns playing a keyboard. One player plays for 15 minutes. Another plays for 22 minutes. A third plays for 34 minutes. And a fourth plays for 40 minutes. How many minutes did they play in all? Convert your answer into hours and minutes.

Bedtime

An astronaut's day is done. It's time to hit the hay. Sleeping in space isn't like sleeping on Earth. In space there is no up or down. An astronaut can't fall out of bed. She floats instead.

At bedtime, astronauts climb into sleeping bags. The bags hang from a wall, ceiling, or seat.

There are two tiny cabins on the ISS. Each cabin has enough room for one sleeping bag. Inside, an astronaut can listen to music or play video games without bothering anyone. She can also look at space through the huge window.

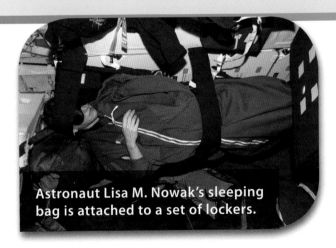
Astronaut Lisa M. Nowak's sleeping bag is attached to a set of lockers.

 What's the Word?

Do you know the expressions "hit the hay" and "hit the sack?" In the early 1900s many people slept on sacks stuffed with hay or straw. So, to hit the sack or to hit the hay meant "to go to bed."

Imagine that an astronaut went to bed at 10:45 P.M. She awoke at 6:30 A.M. How long did the astronaut sleep?

Sunrise, Sunset

People on Earth experience one sunrise and one sunset in a 24-hour period. That's because Earth makes one complete **rotation** in 24 hours. That means it turns all the way around.

But the ISS goes all the way around Earth every 90 minutes. How many times does the ISS orbit Earth in one day?

How many times do astronauts aboard the ISS experience a sunrise?

How many times do they experience a sunset?

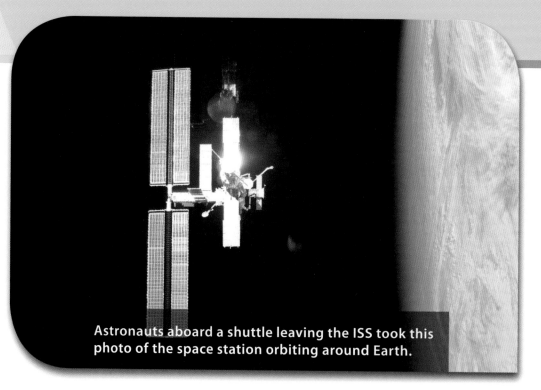

Astronauts aboard a shuttle leaving the ISS took this photo of the space station orbiting around Earth.

Math at Work

Doctors on Earth take care of astronauts both on land and in space. In space, astronauts are hooked to special wires. These wires let doctors on Earth count how many times an astronaut's heart beats each minute. Doctors also measure an astronaut's blood pressure. These measurements tell doctors how hard an astronaut's heart is working. Doctors also measure how many breaths an astronaut takes each minute.

Doctors talk to astronauts every day to check on them. If there's a problem, they take steps to solve it. Some doctors train with astronauts. This helps them understand how space can affect an astronaut's health.

Dr. Josef F. Schmid (right) trains underwater with astronaut José M. Hernandez.

Connecting to History

The ISS is not the first space station. Russia launched a space station called *Mir*. It spent 15 years in orbit. That is a record.

Building *Mir* began in 1986 and lasted for ten years. In time, the station became as big as 10 school buses.

In Russian, the word *mir* has two meanings. One is "peace." The other is "world." The *Mir* space station was a peaceful scientific community. Scientists from around the world worked in labs aboard the station.

Most of the time, three people lived aboard *Mir* for 90 days. How many months is that?

The Soviet Union made this stamp to honor the cooperation of international scientists aboard *Mir*.

 **What's the Word?**

Russian astronauts are called **cosmonauts**. Cosmonaut comes from two Greek words. The first is *kosmos*. That means "universe." The second is *nautes*. That means "sailor."

Scientists traveled to and from *Mir* for 15 years. That was ten years more than the station's builders expected.

In March 2001, Russia sent a rocket to *Mir*. The rocket docked at the space station. It fired its engines to slow down *Mir* and to push it in the opposite direction.

This picture shows the U.S. space shuttle *Atlantis* docked at *Mir* space station.

This started the station's fall back to Earth. Most of the station burned up as it moved through Earth's **atmosphere**. But 20 tons of broken pieces fell into the ocean. Those pieces hit water in the South Pacific Ocean on March 23, 2011. *Mir* is the largest space ship to ever come back from space.

Say that broken parts of *Mir* crossed the sky above Japan at 3:30 P.M. *Mir* had entered the atmosphere at 2:07 P.M. How much time elapsed by the time people in Japan could see *Mir's* parts?

Astronaut Joe Acaba and two other astronauts blasted off from a space center in Kazakhstan on May 14, 2012. Their space craft launched at 11:01 P.M. It docked with the ISS three days later, at 10:45 P.M.

How many hours elapsed between take-off and docking? How will you find the answer?

Idea 1: You can **move the hands of a clock** to find the elapsed time. It might be difficult to keep up with how much time has passed on a clock.

Idea 2: You can **write and solve a subtraction problem**. You might need to write and solve several problems. This could take time.

Idea 3: You can use a **number line diagram**. Two full days elapse between Monday and Wednesday. That's 48 hours. Draw a number line to find how much time elapsed between Wednesday and Thursday. Then, add to find out how much total time elapsed.

Maybe someday you will become an astronaut and live aboard the ISS. But for now, you'll have to look for it up in the sky!

WHAT COMES NEXT?

If you were designing a space station, what would it look like? Create a list of the equipment you might need on your space station. Do you think you would need computers? How about a camera and a telescope? What's the best source of energy for the machines you use?

Where would you like to send your space station? You can orbit Earth or a different planet. What would you want to find out about each planet? Share your ideas with a friend or an adult.

How many people do you want your space station to hold? How much water should you bring? How much food? How long do you think you will be aboard the station?

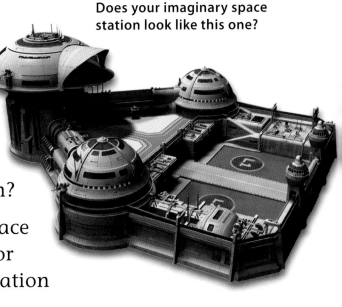

Does your imaginary space station look like this one?

Make a model of your space station. You can draw it or build it. Use your imagination and have fun!

GLOSSARY

atmosphere: the mass of air surrounding Earth.

convert: to change units into units of another kind.

cosmonauts: astronauts of Russia and the former Soviet Union.

dehydrated: foods preserved by the removal of moisture.

elapsed time: time that has passed.

equivalent: being the same, or nearly the same in value.

force: a push or pull exerted on an object.

gravity: a force that pulls two objects to each other.

mass: the amount of matter in an object.

microgravity: small amounts of gravity, also known as weightlessness.

mold: tiny living things that live in the soil and on dead plants and animals.

number line diagram: a diagram that uses points on a line to represent numbers.

orbit: to move in a path around another object.

proteins: the building blocks of all life; materials in living things that serve many purposes, such as delivering oxygen to cells.

regrouping chart: a chart that changes units into other units before adding or subtracting.

resistance: a force that slows down another force.

rotation: to move in a complete circle around a point.

schedule: a plan of work, showing tasks that are to be carried out at given times.

solar cells: objects that change solar, or light energy, into electrical energy.

space shuttle: a rocket-launched spacecraft that orbits Earth and is able to land like an airplane.

subtraction problem: a problem that finds the difference between two numbers.

treadmill: an exercise machine with a revolving belt on which someone can walk, jog, or run.

FURTHER READING

NONFICTION

Exploring the International Space Station, by Laura Hamilton Waxman, Lerner Classroom, 2011

13 Planets: The Latest View of the Solar System, by David A. Aguilar, National Geographic Kids, 2011

Life on a Space Station, by Andre Einspruch, PowerKids Press, 2012

FICTION

Space Station Rat, by Michael J. Daley, Holiday House, 2008

ADDITIONAL NOTES

The page references below provide answers to questions asked throughout the book. Questions whose answers will vary are not addressed.

Page 13: a. Elapsed Time: 1 hour, 15 minutes; b. Start Time: 10:30 A.M.; c. Start Time: 2:05 P.M.; d. Elapsed Time: 4 hours, 55 minutes; e. End Time: 5:35 P.M.; Did You Know: $42\frac{1}{2}$ days

Page 15: 12 hours, 35 minutes

Page 17: 8 hours, 10 minutes

Page 19: 4:35 P.M.

Page 21: His workout ended at 1:45 P.M. and he exercised for a total of 1 hour and 45 minutes.

Page 22: They played for 111 minutes or 1 hour and 51 minutes.

Page 23: The astronaut slept for 7 hours and 45 minutes.

Page 24: ISS orbits Earth 16 times a day; 16 sunrises, 16 sunsets

Page 26: about 3 months

Page 27: 1 hour, 23 minutes

Page 28: 71 hours and 44 minutes

INDEX

CONTENT CONSULTANT

David T. Hughes
David is an experienced mathematics teacher, writer, presenter, and adviser. He serves as a consultant for the Partnership for Assessment of Readiness for College and Careers. David has also worked as the Senior Program Coordinator for the Charles A. Dana Center at The University of Texas at Austin and was an editor and contributor for the *Mathematics Standards in the Classroom* series.